BOB BARNER

DINOSAURS DEPART

A *Start Smart* Math Book

ROOSTER BOOKS

BANTAM DOUBLEDAY DELL
New York Toronto London Sydney Auckland

DINOSAURS DEPART
A Rooster Book/June 1996

For information, address Bantam Doubleday Dell Books for Young Readers.

Library of Congress Cataloging-in-Publication Data
Barner, Bob.
Dinosaurs depart / Bob Barner.
p. cm.—(A start smart math book ; 4)
ISBN 0-553-37584-9 (pbk.)
1. Subtraction—Juvenile literature.
[1. Subtraction. 2. Dinosaurs. 3. Mathematical recreations.]
I. Title. II. Series: Barner, Bob. Start smart math book ; 4.
QA115.B36 1996 95-10630
513.2' 12—dc20 CIP AC

Published simultaneously in the United States and Canada
Rooster Books are published by Bantam Doubleday Dell Books for Young Readers,
a division of Bantam Doubleday Dell Publishing Group, Inc., 1540 Broadway,
New York, New York 10036.
10 9 8 7 6 5 4 3 2 1
Printed in USA

for Paul and Darlene,
who frequently depart

NOTES ABOUT THIS BOOK

Problem solving with subtraction is the central focus of this book. By using a problem-solving approach to the study of mathematics, children learn to investigate and understand mathematical content; they also develop inquiring minds. The exploration of subtraction is central to understanding mathematics. Children who are given an opportunity to model and discuss a variety of problem situations are better able to develop a real understanding of subtraction. This provides a framework for mental and written computation.

As you read the story, talk about the dinosaurs and their problem — how to get everyone on one bicycle. Talk about how one of the dinosaurs uses a problem-solving strategy—logical thinking—to solve their problem. Help children see that, while five of the dinosaurs try unsuccessfully to pile onto a bicycle with one seat, one of the dinosaurs quietly thinks about the problem and finds a logical solution—a bicycle with six seats. Encourage children to explain the problem in their own words. Discuss the fact that there are always six dinosaurs, even though the quantity of dinosaurs at home and on the bicycle changes.

Remind children that the number word for none is "zero." Be enthusiastic. Help children enjoy this book. Success, praise, and enjoyment are all crucial for the development of a positive attitude toward math. Remember that for younger children, the learning experience is as important as a "right" or "wrong" answer. Of course, there is no replacement for your own enthusiasm, praise, and enjoyment.

JO DENNIS
Math Education Consultant

On Friday six dinosaurs are making art.

But when school is out, they'll all depart.

When they hear the final bell,

six dinosaurs all cheer and yell.

Now that summer has begun,

six dinosaurs are ready for fun.

But on Saturday they're in a pickle.
Six can't ride on one bicycle.

On Sunday one leaves, quiet as a mouse.
How many dinosaurs are left at the house?

On Monday two dinosaurs sneak away.
Now how many will have to stay?

On Tuesday three dinosaurs get their wishes.
How many are left to do the dishes?

On Wednesday four are off to the sea.
How many are left where six used to be?

On Thursday five dinosaurs ride away.
How many then decide to stay?

On Friday one has a big surprise.

The other five can't believe their eyes!

Six dinosaurs happily pedal to the sea.
How many are left where six used to be?

And now as you can plainly see,

six dinosaurs all shout, “Whoopee!”

SUBTRACTING FROM SIX.

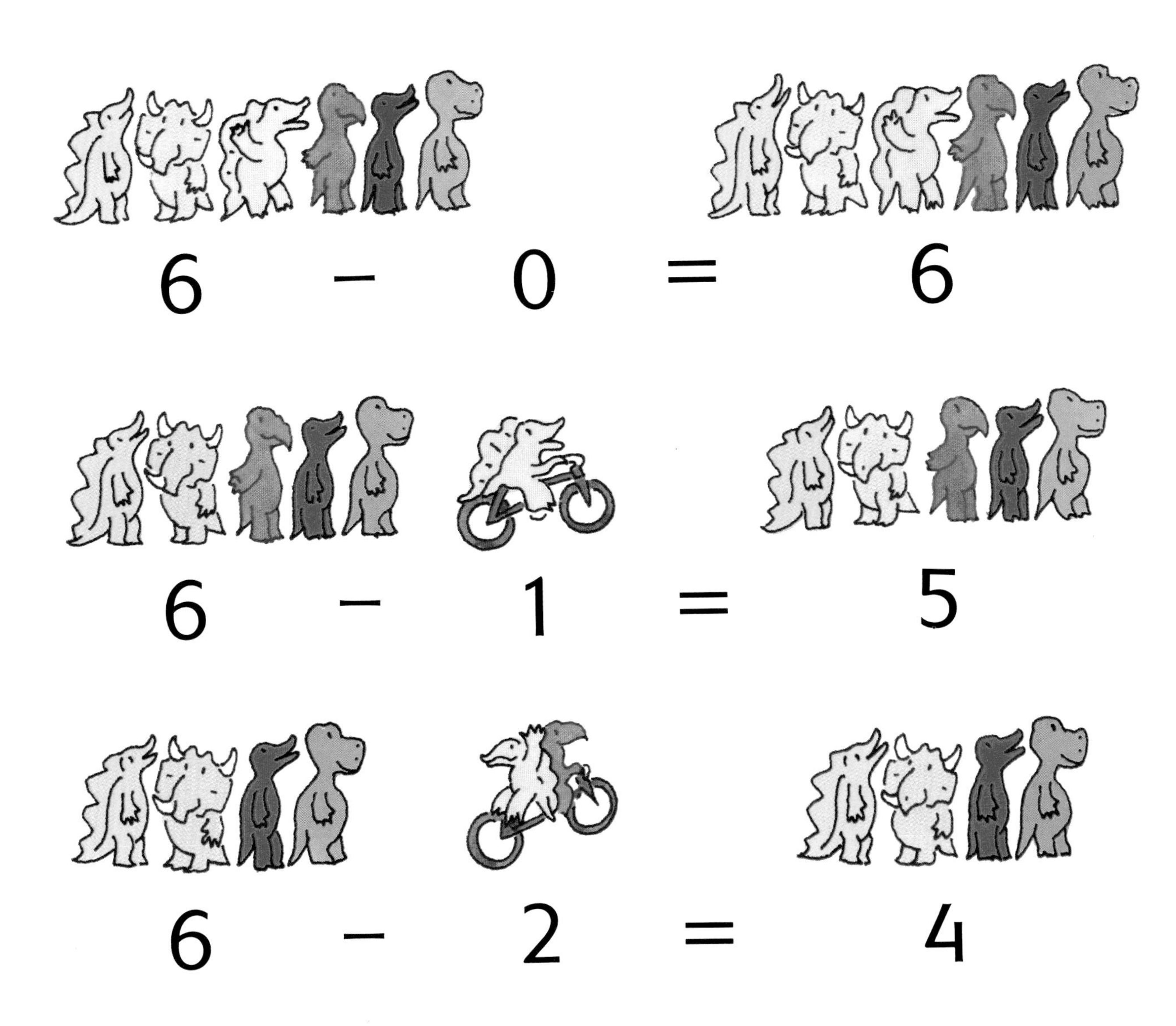

$6 - 0 = 6$

$6 - 1 = 5$

$6 - 2 = 4$

6 − 3 = 3
6 − 4 = 2
6 − 5 = 1
6 − 6 = 0

ACTIVITIES WITH PUNCH-OUT MANIPULATIVES

by Jo Dennis, Math Education Consultant

1. Children can use six of the punch-out dinosaurs and the workmat of the house and the bicycle to act out the story. They can make up a subtraction story problem for each fact and challenge others to solve it. For example: "Six dinosaurs live in the house. One dinosaur rides away. How many dinosaurs are left in the house?"

2. Children can use six of the punch-out dinosaurs and the workmat of the house and the bicycle to make up some addition story problems for others to solve. For example: "There are three dinosaurs in the house and three dinosaurs on the bicycle. How many dinosaurs are there altogether?"

3. Children can use the workmat and different quantities of dinosaurs—starting with six, then five, then four, and so on, to none—to make as many subtraction facts as possible. As they explore with each quantity of dinosaurs, children can make a list of all the possible facts. For example:

6	**5**	**4**	**3**	**2**	**1**	**0**
6 – 0	6 – 1	6 – 2	6 – 3	6 – 4	6 – 5	6 – 6
	5 – 0	5 – 1	5 – 2	5 – 3	5 – 4	5 – 5
		4 – 0	4 – 1	4 – 2	4 – 3	4 – 4
			3 – 0	3 – 1	3 – 2	3 – 3
				2 – 0	2 – 1	2 – 2
					1 – 0	1 – 1
						0 – 0